Juan Gabriel de Albuquerque Ramos

Analysing Energy Bills: Solutions for overrun fines

Juan Gabriel de Albuquerque Ramos

Analysing Energy Bills: Solutions for overrun fines

Case study of the Federal University of Amazonas

ScienciaScripts

Cover image: www.ingimage.com

This book is a translation from the original published under ISBN 978-3-330-76352-4.

Publisher:
Sciencia Scripts
is a trademark of
Dodo Books Indian Ocean Ltd. and OmniScriptum S.R.L publishing group

120 High Road, East Finchley, London, N2 9ED, United Kingdom
Str. Armeneasca 28/1, office 1, Chisinau MD-2012, Republic of Moldova, Europe
Managing Directors: Ieva Konstantinova, Victoria Ursu
info@omniscriptum.com

Printed at: see last page
ISBN: 978-620-8-40518-2

Juan Gabriel de Albuquerque Ramos

Analysing Energy Bills: Solutions for overrun fines

Case study of the Federal University of Amazonas

ScienciaScripts

Imprint

Cover image: www.ingimage.com

This book is a translation from the original published under ISBN 978-3-330-76352-4.

Publisher:
Sciencia Scripts
is a trademark of
Dodo Books Indian Ocean Ltd. and OmniScriptum S.R.L publishing group

120 High Road, East Finchley, London, N2 9ED, United Kingdom
Str. Armeneasca 28/1, office 1, Chisinau MD-2012, Republic of Moldova, Europe
Managing Directors: Ieva Konstantinova, Victoria Ursu
info@omniscriptum.com

Printed at: see last page
ISBN: 978-620-8-40518-2

ACKNOWLEDGEMENTS

I would firstly like to thank God, the author, the beginning and the end of all human wisdom. I also thank my family who have always helped me to be a good person. I also thank the Federal University of Amazonas, the professors at the Faculty of Technology and, finally, my wife Camila who always makes me happy.

Summary

CHAPTER 1

INTRODUCTION

Among the many items that can be commercialised today, the one with the most particular characteristics is electricity. Yes, in addition to the various definitions postulated in physics textbooks, there is the definition that energy is a commodity that can be commercialised.

Electricity also represents a fundamental item for the maintenance and functioning of the model of life currently employed. As such, energy trade needs to be properly studied and regulated so that it can take place in the safest and most harmonious way possible.

The study of energy pricing deals with the evaluation of electricity consumption and its economic impact. From the point of view of a large consumer, a well-designed study can reduce electricity costs without resulting in a loss of productivity.

In the course of this work, a problem faced today by the Federal University of Amazonas will be shown, which is the payment of fines for exceeding the contracted demand, a fact that has generated a high monthly energy bill for the University.

1.1. OBJECTIVES

The aim of this book is to present a case study of the electricity bills of the Federal University of Amazonas. The study is designed to create a standard that can be applied in a program that can analyse the history of bills and generate an optimal demand value. With this, the aim is to focus on equating the problem to the various possible situations that would improve the situation of a consumer unit, in this specific case the Senator Arthur Virgílio Filho Campus.

With the aim of creating a standard to be used in the coming years, deals not only with the charging model previously used, but also with the new model that was implemented in 2014.

1.2. BOOK ORGANISATION

This work is organised into 11 chapters, the first seven of which are for a literature review and contextualisation on the subject of energy tariffs.

The eighth chapter deals with the case study that was carried out by analysing the 12-month history of UFAM's electricity bills, from June 2012 to May 2013. This case study proposes changes to UFAM's demand contracts.

Chapter 9 then deals with the author's final considerations and suggestions for generating improvements in the university's electricity consumption.

Chapter 10 is a conclusion, and Chapter 11 deals with bibliographical references.

CHAPTER 2

HISTORY OF ENERGY PRICING IN BRAZIL

To understand how the electricity pricing model currently used in Brazil works, you need to know that it is the product of more than 100 years of evolution. This development was generated by the relationship between the political, economic and social development of Brazil and the world.

According to a study carried out in [1] and [2], the history of electricity pricing in Brazil is as follows: it began in 1901, with the advent of the Canadian company called Light Serviços de Eletricidade S.A. During this time, also known as the Old Republic, the regulation of tariffs was characterised by the use of the Golden Clause in contracts with Light. The tariffs that were applied at this time were created in order to encourage foreign distributors and favour their development and expansion, with the result of increasing the country's attractiveness to foreign investment and cultural promotion of the use of electricity.

Over time, the landscape of Brazilian politics changed and the Brazilian government model became more involved in energy matters. The government began to have more influence over the processes of regulating and supervising energy metering. In 1934, it changed the tariff framework with the publication of the Water Code. This repealed the Golden Clause and a new tariff model was created. In it, the energy tariff is now measured by the cost of the service, while the distribution company must receive a guaranteed 10% return on the assets mobilised. The new model was complemented by Decree 41.019/57, which defined the creation of consumer classes, and Decree 62.724/68, which led to the current binomial structure. The decree also defined the apportionment of service costs equally between all consumer groups.

Over time, electricity pricing began to be used as an economic instrument. A good example is Decree No. 1,383 of 1974. This document, which created the Global Guarantee Reserve, was used to transfer resources from more profitable companies to less profitable ones. This system, considered by many to be inappropriate, resulted in an increase in the foreign debt of some companies and created a financial crisis in the electricity sector.

As indicated in the study [1], in 1979 the Ministry of Mines and Energy drew up a document entitled "Modelo Energético Brasileiro" (Brazilian Energy Model). This document, based on a series of energy studies, showed that measures had to be taken to get out of the crisis in the electricity sector. Among the measures was restructuring the tariff structure in order to properly distribute the costs associated with generation, transmission and distribution.

After this stage, the issue of energy tariffs began to be addressed more directly. The new tariff system made it possible to create an economic signal for consumers, encouraging consumption during periods of lower demand. In 1982, the first ordinances were published, defining the hourly-seasonal modalities for large consumers and planning their application to low-voltage consumers. The guaranteed return system ceased to be applied with the use of standardised tariffs in 1993. To this end, changes were made to the legislation that determined the use of the **price cap** concept.

Created in 2001, the Committee to Revitalise the Energy Sector Model decided to re-level the tariffs by adopting the calculation of energy tariffs (TE) and charges for the use of the distribution system (TUSD).

These are the tariffs that are applied to the tariff structure that is currently in force.

In order to keep the tariff structure up to date, the National Electricity Agency is already planning to create a new tariff model. Due to be implemented in 2014 [3], the new model addresses variations in the cost of energy throughout the day and year, with a new proposal that combines past experience with the technology available today.

CHAPTER 3

THE NATIONAL ELECTRICITY AGENCY

Created by Law No. 9.427 [4], the National Electric Energy Agency (ANEEL) is a regulatory agency linked to the Ministry of Mines and Energy whose aim is to regulate and supervise the production, transmission and distribution of electricity, in accordance with the guidelines of the Brazilian government.

As can be seen historically, for the energy market to function fairly and efficiently, the pricing system applied to the sale of electricity needs to be very well planned, documented and regulated.

Therefore, it is concluded that the study presented in this book must comply with the rules and definitions established by ANEEL, and the documents containing the definitions and tariffs presented in this work can be easily found at the following e-mail address: http://www3.aneel.gov.br/pesquisadigit.htm.

CHAPTER 4

CONSUMERS

When it comes to electricity consumption, the definition of consumer differs somewhat from that applied in everyday life. According to ANEEL [5], an energy consumer is defined as a natural or legal person who requests the supply of energy or the use of the electricity system from the distributor, assuming the obligations arising from this service.

Although not everyone is charged in the same way, energy consumers are responsible for charges related not only to their consumption but also to the availability of supply. This difference will become clearer with the concepts of consumption and demand that will be presented below.

For the purposes of applying energy tariffs, consumers are divided by ANEEL [4] into two tariff groups called group A and group B.

4.1. GROUP B

Group B covers the so-called small consumers, consumer units whose supply voltage is less than 2.3kV. Thanks to the variability of consumption patterns in this group, consumers are subdivided into four subgroups called B1, B2, B3 and B4, which represent residential, rural, other classes and public lighting respectively. This subdivision allows tariffs and discounts with different values to be applied to consumers with different needs

The tariff model applied to these consumers is the simplest of all. It only takes into account the registered monthly consumption value of the consumer unit, which characterises the monomial type tariff.

4.2. GROUP A

Group A is made up of consumers who have a supply voltage of 2.3 kV or more, or who are served from an underground distribution system at secondary voltage, which is a voltage of less than 2.3 kV. Energy consumption in this group is billed using the binomial tariff system. This system takes into account the registered consumption and billable demand values, hence the term binomial.

In the same way that group B is subdivided into smaller groups, ANEEL defined that group A should be subdivided into six other tariff subgroups, which are listed below.

- A1 - Voltage level equal to or greater than 230kV.
- A2 - Voltage level from 88 to 138kV.
- A3 - Voltage level 69kV.
- A3a-Voltage level 30 to 44kV.
- A4 - Voltage level from 2.3 to 25kV.
- AS - Underground system.

In order to respect the characteristics of consumers and encourage the best use of energy and the electricity system, ANEEL has defined three tariff types that can be chosen by consumers according to their subgroup.

CHAPTER 5

CURRENT TARIFF MODALITIES

ANEEL [5] defines the term tariff modality as "a set of tariffs applicable to the components of electricity consumption and active power demand". Currently, the tariff structure used in Brazil utilises four types of tariff:

- Conventional monomial
- Conventional binomial (Conventional)
- Green seasonal schedule (Green)
- Blue seasonal (Blue)

Each modality aims to charge consumers belonging to a certain sub-group as fairly as possible. This is done by applying a calculation specific to each modality that determines the Partial Value of the Bill (VPF) of energy that must be paid to the concessionaire. The term VPF is used because the final billing also includes the costs associated with fines and chargeable services such as: inspections, reconnections, meter measurements, etc. The values of these services are approved annually along with the tariffs.

The first model, conventional monomial, is applied only to group B consumers, while the following three are available specifically for group A consumers according to Table 1 [6]:

It is important to emphasise that although the calculation of a modality is the same for different consumers, the tariffs used in the calculations are calculated specifically for each consumer subgroup of each distributor.

Table 1 - Availability of group A modalities

<table>
<tr><th rowspan="2">Tariff subgroup</th><th colspan="3">Tariff Modality</th></tr>
<tr><th>Conventional</th><th>Green</th><th>Blue</th></tr>
<tr><td>A1</td><td rowspan="4">Impeded</td><td rowspan="4">Impeded</td><td rowspan="4">Compulsory for any contracted demand value</td></tr>
<tr><td>A2</td></tr>
<tr><td>A3</td></tr>
<tr><td></td></tr>
</table>

A3a	Available for contracts under 300 kW	Available for I[1] I[1] 1 any contracted demand value	Available for any contracted demand
A4			
AS			

SOURCE: Elektro, Elektro Energy Efficiency Manual, p.37.

In order to present the simplest tariff models, it is important to define four concepts that form the basis of every tariff model. These are

- Consumption
- Demand
- Contracted Demand.
- Billable Demand

5.1. BASIC CONCEPTS OF CONSUMPTION AND DEMAND

5.1.1. Consumption

One of the most important parameters in calculating tariffs. It represents the use of active power during a given time interval. For the utility company, this quantity, usually expressed in kWh, represents the amount of electricity that the consumer buys from the distribution system to carry out a given job.

5.1.2. Demand

Currently, the demand value is used specifically for the modalities that work with consumers in tariff group A. As these consumers affect the electricity system much more, individually speaking, the meters used are more complete than those used for consumers in group B, who only measure consumption. The demand value is obtained from the recorded consumption data that is taken periodically throughout the day.

Demand is defined as follows [2]: "It is the average of the active or reactive electrical powers requested from the electrical system by the portion of the installed load in operation at the consumer unit, during a specified time interval." Since 1968, the time interval defined for use in Brazil has been 15 minutes [7].

5.1.3. Contracted Demand

As its name implies, contracted demand is a fixed amount of demand that a consumer contracts from a utility. This value has three objectives: the first is to guarantee the consumer that they will always have at their disposal the demand value they have set themselves. The second is to guarantee the distributor a forecast and guarantee that the demand will be met. The contracted demand data is used in the process of planning and expanding the distribution network.

The third objective of contracted demand is to encourage conscientious electricity consumption. Once the contracted demand amount is charged, consumers tend to improve their consumption pattern in order to make better use of the investment they have made.

Proof of the importance of respecting the value of the contracted demand for the electricity system is the value of the overrun fine. This amount, which is around three times the demand, makes it very difficult for a consumer to have a profile that exceeds the contracted demand and shows a lower value than if they respected the rules.

5.2. CONVENTIONAL MODE

5.2.1. Monomyth

As previously mentioned, the conventional monomial modality is the simplest of all to calculate. The VPF is given according to Equation 1.

$$VPF = \frac{CF.TC}{(1 - ICMS)} \quad (1)$$

Where:

CF - Recorded consumption value;

TC - Current consumption tariff;

ICMS - Goods Circulation Tax Index;

Although this modality was created to specifically cater for Group B consumers, it can also be applied to Group A consumers with the following characteristics [8]:

- Located in holiday or tourism areas that operate hotel or hostel activities, regardless of the installed load.
- Consumers whose installed power in transformers is less than 112.5 kVA.

- Permanent installations for sports or agricultural parks, provided that the power installed in the reflectors is equal to or greater than 2/3 of the consumer unit's installed load.

5.2.2. Binomial

According to the literature [2], the VPF of the conventional binomial model is calculated taking into account the consumer's consumption and demand values according to Equation 2.

$$\frac{(CF.TC + DF.TD)}{(1 - ICMS)} \quad (2)$$

Where:

CF - Recorded consumption value;

TC - Current consumption tariff;

DF - Invoiced demand value;

TD - Current demand tariff

ICMS - Goods Circulation Tax Index;

5.3. HOURLY MODALITIES

Seasonal rates differ from other rates in two respects. The first is that the tariffs applied in the VPF calculations vary according to the time of year. The second is that the VPF calculation also takes into account the consumption and demand values recorded throughout the day.

As can be seen in Figure 1 [9], 81.9 per cent of the electricity supplied in Brazil comes from hydroelectric sources. As a result, the availability of energy depends on the amount of water contained in the reservoirs, meaning that the availability of energy is linked to climatic variations.

One of the ways to guarantee the security of energy availability is to produce energy from different energy sources. In Brazil, thermoelectric plants have been chosen to ensure availability. These plants come into operation when the level of the reservoirs falls beyond a minimum threshold, and the electricity system starts using more expensive energy, which results in an increase in the overall price of electricity.

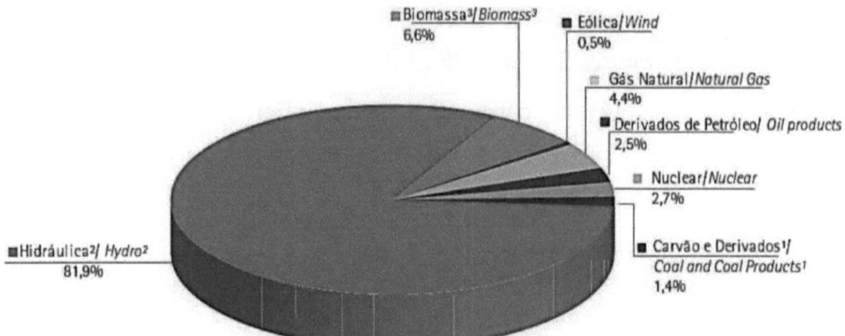

Figure 1 - Domestic electricity supply by source

Source: Ministry of Mines and Energy, National Energy Balance, 2012, p.16

To compensate for the variation in the cost of energy and to encourage energy saving during the driest periods of the year, two periods were defined: the dry period and the wet period. Specific tariffs have been established for each period, which are used in the hourly and seasonal modalities.

From the distributor's financial point of view, the best consumption pattern would be a constant. In this way, the distribution system would never be idle and could operate at its best. However, this is not the case with actual consumption patterns. An example of the typical load curve seen by a distributor is illustrated [10] by Figure 2, which shows the curve expected by the National Electricity System Operator for the Southeast-Central West subsystem, São Paulo area, in 2009 (ONS, 2007).

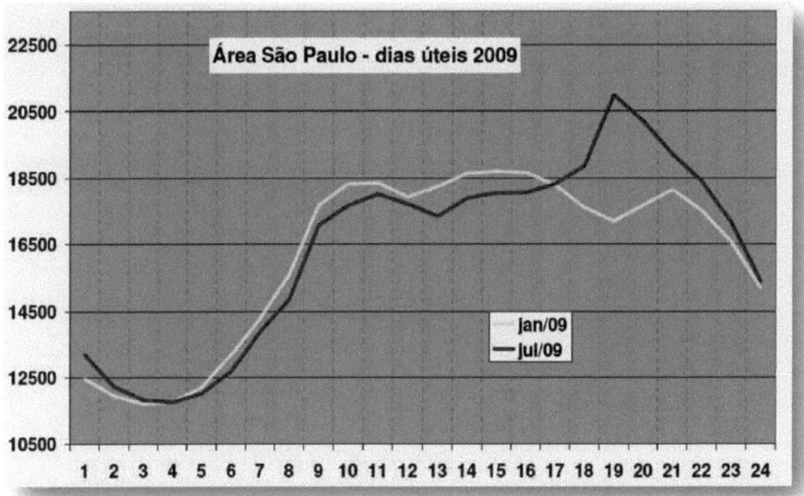

Figure 2 - São Paulo Area Load Curve

Source: ONS, Consolidation of the load for 2008-2010, 2007, p.34

As is to be expected, except in very rare cases, consumers have a cyclical consumption pattern that varies over time. This inconstant consumption causes the distribution system to experience moments of idleness and consumption peaks that are undesirable in the eyes of the distributor, as they generate costs and wasted investment.

In order to alleviate this problem, the peak and off-peak periods were defined, used in the hourly-seasonal modalities, with the aim of influencing consumers' consumption patterns so that they make better use of the electricity system. These concepts, as well as the dry and wet periods regulated by ANEEL [5], are described below.

5.3.1. Period concepts

5.3.1.1 Peak Period

Also known as the peak tariff period, it is characterised by tariffs that aim to shift consumption to other periods of the day. It is defined as a time interval consisting of three consecutive daily hours defined by the distributor considering the load curve of its electrical system. This period, which is applied from Monday to Friday (except for the public holidays shown in Table 2 [11]), is normally defined as the period between 6pm and 9pm, which is when the electricity system works at its highest load.

Table 2- Days when the peak period is not applied

Day and month	**National holidays**
01 January	Universal Confraternisation
21st April	I iradentes
01 May	Labour Day
0/September	Independence
12th October	Our Lady of Aparecida
02 November	All Souls
15th November	Proclamation of the Republic
25th December	Christmas

Source: Escelsa, Manual do cliente horossazonal, 2004, p25

5.3.1.1 Off-peak period

The off-peak period, also known as the off-peak tariff period, is defined as the period made up of all the daily hours that are consecutive and complementary to the hours that make up the peak period. Unlike what happens during the peak period, the tariffs applied to the consumption and demand values recorded during this period are lighter, which gives consumers greater freedom to use energy in the way that suits them best, while still respecting the contracted demand value agreed with the utility.

The weekday and weekend tariffs are exemplified by Figure 3 [14] in which the blocks with the letter "P" represent the hours that make up the peak period, and the letters "FP" represent the hours that make up the off-peak period.

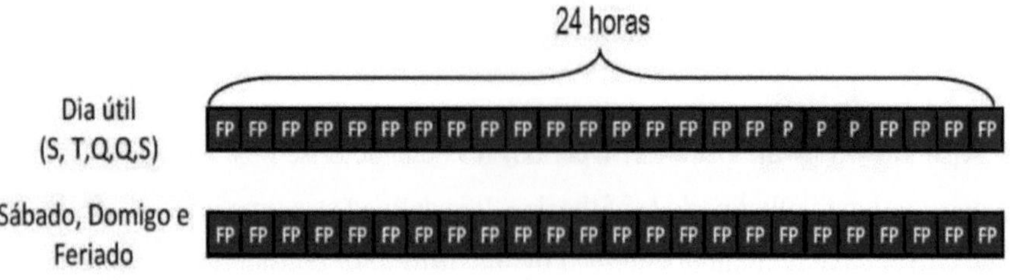

Figure 3 - Definition of tariff stations.

Source: ANEEL, Technical note n°311/2011, 2011, p14

5.3.1.3 . Dry Period

The dry period is defined as the period of five consecutive billing cycles, from December of one year to April of the following year. In a similar way to peak hours, demand and consumption recorded during the dry period is charged using more expensive tariffs than those used for consumption during the wet period.

5.3.1.4 . Wet Period

The wet period is defined as the period of seven consecutive billing cycles, from May to November. As such, it represents the complementary period to the dry one, and in it lighter tariffs are applied to the recorded demand and consumption values.

5.3.2. Green Seasonal Mode

A method characterised by the application of differentiated electricity consumption tariffs, according to the time of day and period of the year. There are also differentiated demand tariffs according to the time of year, as well as a single power demand tariff.

The VPF for this modality is calculated using Equations (3) and (4) [2]. The first presents the equation for the values recorded in the dry period, and the second the equation for the wet period.

$$VPF = \frac{CF_{fs}.TC_{fs} + CF_{ps}.TC_{ps} + DF.TD}{(1 - ICMS)} \quad (3)$$

$$VPF = \frac{CF_{fu}.TC_{fu} + CF_{pu}.TC_{pu} + DF.TD}{(1 - ICMS)} \quad (4)$$

Where indices mean:

f - off-peak

p - Peak times

s - dry season

u - wet season

5.3.3. Blue Seasonal Mode

Mode characterised by the application of differentiated electricity consumption tariffs, according to the hours of use of the day and the periods of the year. The tariffs related to billed demand are also differentiated according to the hours of use of the day and the periods of the year.The VPF of the blue seasonal hour mode is calculated using Equations 5 and 6 [2] :

$$VPF = \frac{CF_{fs}.TC_{fs} + CF_{ps}.TC_{ps} + DF_{f}.TF_{f} + DF_{p}.TD_{p}}{(1 - ICMS)} \qquad (5)$$

$$VPF = \frac{CF_{fu}.TC_{fu} + CF_{pu}.TC_{pu} + DF_{f}.TF_{f} + DF_{p}.TD_{p}}{(1 - ICMS)} \qquad (6)$$

A special feature of this modality is that it is the only one that uses two different contracted demand values, one for the peak period and one for the off-peak period.

CHAPTER 6

FINES

Although the VPF defined by the modalities does not take fines into account, it is important to consider them in order to know the limitations that must be taken into account when minimising the value of the VPF. The main fines are for transgressing the power factor and for exceeding the contracted demand.

6.1. POWER FACTOR

According to ANEEL [5] the power factor is defined by Equation 7:

$$fp = \frac{P}{\sqrt{P^2 + Q^2}} \qquad (7)$$

Where:

fa = power factor;

P = active power

Q = reactive power

Group A consumers are obliged to have their power factor checked by the distributor using a permanent meter. Measuring the power factor of group B consumers is optional.

ANEEL [5] also defines the minimum limit for consumer units as 0.92. If a power factor of less than 0.92 is recorded (inductive or capacitive), the consumer must pay the fine established by Equations 8 and 9:

$$E_{RE} = \sum_{T=1}^{n} \left[EEAM_T . \left(\frac{f_r}{f_t} - 1\right) \right] . VR_{ERE} \qquad (8)$$

$$D_{RE}(p) = \left[MAX_{T=1}^{n} \left(PAM . \frac{f_r}{f_t} \right) - PAF(p) \right] . VR_{ERE} \qquad (9)$$

Equation 10 below shows how the electricity bill is adjusted for a power factor below that determined by ANEEL resolution 456/2000, which is 0.92:

$$Acréscimo = Valor\ da\ fatura\ x\ \left[\frac{0{,}92}{fator\ de\ potencia\ medido} - 1\right] \quad (10)$$

For more information on the constants used, see Normative Resolution No. 414.

6.2. EXCEEDING THE CONTRACTED DEMAND

As explained above, the contracted demand represents the maximum amount of demand that the consumer undertakes to use during the measurement cycle. The fine for exceeding demand is applied when the registered active power demand exceeds the contracted demand value by more than 5 per cent. In this case, the consumer will pay for the use of the demand and the fine established by Equation 11 [5]:

$$D_{ultrapassagem} = [PAM(p) - PAC(p)].2.VR_{dult}(p) \quad (11)$$

Where:

PAM(p) = Active power measured at each time station "p";

PAC(p) = Active power contracted at each time station "p";

VR_{dult} = Reference value equivalent to power demand tariffs.

CHAPTER 7

TARIFF

The construction of the tariffs used in the VPF calculations for each modality is the result of an overly complex process that is beyond the scope of this work. However, their basic construction is important for understanding how the new model that is being implemented works.

7.1. TARIFF DEFINITION

According to ANEEL, tariffs are defined as an established monetary value, set in Reais per unit of active electricity or active power demand. Their purpose is to maintain harmony in the energy market in order to guarantee a fair price for consumers and an adequate financial return for energy distributors.

In order to treat both consumers and distributors fairly, the value of each tariff varies according to the physical and economic structure of the distributor, the associated modality and the tariff station. Since distributors and the effect of tariffs on the market are always changing over time, tariffs are updated annually considering a series of cost functions that are obtained from a series of distributor data.

The consumption (TC) and demand (TD) tariffs used to calculate the VPFs of the tariff types are made up of two basic components: the Distribution System Use Tariff (TUSD) and the Energy Tariff (TE). In general terms, this relationship can be seen in Equations 12 and 13.

$$TC = TUSD + TE \quad (12)$$

$$TD = TUSD \quad (13)$$

The relationship shown by equations 12 and 13 can be easily verified by studying Table 3, which shows an excerpt from homologatory resolution no. 1072 that homologates the electricity supply tariffs and the tariffs for the use of distribution systems for Bandeirante Energia S.A. - Bandeirante.

Table 3- Tariffs for the Bandeirante distributor's Blue mode

BLUE SEASONAL TARIFF	TABLE B					
	TUSD + TE		TUSD		TE	
	DEMAND (R$/Kw)		DEMAND (R$/Kw)		DEMAND (R$/Kw)	
SUBGROUP	TIP	F. PONTA	TIP	F. PONTA	TIP	F. PONTA
A2 (88 to 138 Kv)	19.12	2,56	19,12	2,56	0,00	0,00
A3a (30 to 44 Kv)	22,45	4,66	22.45	4,66	0,00	0,00
A4 (2.3 to 25 Kv)	30,56	7,35	30,56	7,35	0.00	0,00

BLUE HOURLY SEASONAL TARIFF	TABLE C											
	TUSD+ TE				TUSD				TE			
	ENERGY (RS/MWh)				ENERGY (RS/MWh)				ENERGY (RS/MWh)			
	TIP		F. PONTA		TIP		F. PONTA		TIP		F. PONTA	
SUBGROUP	DRY	WET	DRY	WET	DRY	WET	DRY	WET	DRY	WET	DRY	WET
AI (230Kvou ma is)	250,73	227,39	158,84	145,30	31,83	31,83	31,83	31,83	218,90	195,56	127,01	113,47
A2 (88 to 138 Kv)	250,73	227,39	158,84	145,30	31,83	31,83	31,83	31,83	218,90	195,56	127,01	113.47
A3a (30 to 44 Kv)	250,73	227,39	158.84	145,30	31,83	31,83	31,83	31.83	218.90	195,56	127,01	113.47
A4 (2.3 to 25 Kv)	250,73	227,39	158,84	145,30	31,83	31,83	31.83	31,83	218,90	195,56	127,01	113,47

Source: ANEEL, Homologatory Resolution No. 1. 072, 2010, p.3

As shown in Table 3, consumption and demand tariffs are presented as the sum of the TUSD and TE for each subgroup.

7.2. DISTRIBUTION SYSTEM USE TARIFF (TUSD)

The distribution tariff is divided into two parts, called "Parcel A" and "Parcel B".

Parcel A is made up of non-manageable costs in which the concessionaire only charges the end consumer the amounts necessary to reimburse the amount spent. The components of Parcel A can be grouped into Energy Purchase, Sectoral Charges and Transmission Charges.

Power Purchase is the electricity purchased from generating companies through auctions organised by ANEEL and operated by the CCEE, power purchase and sale contracts signed directly with the generators and the energy compulsorily purchased from the Itaipu Hydroelectric Power Plant.

To serve consumers in its concession area, the distributor buys energy from different generating companies under different conditions, depending on market growth and the region in which it is located.

Expenditure on the purchase of energy for resale is the non-manageable cost item of significant relative weight for distribution utilities. energy can be acquired by the distribution company in basically two ways:

Through the energy auctions mentioned above or through Long or Short Term Bilateral Contracts, which are energy purchases made by distribution companies to supplement the energy needed to fully serve their consumer market, carried out through long or short term bilateral contracts, based on the legal commercialisation mechanisms in force.

One exception is the distribution companies located in the South, Southeast and Centre-West regions of Brazil, which are legally obliged to pay a share of the costs of the electricity produced by Itaipu and destined for the country.

Sectoral Charges are amounts paid by consumers in their electricity bills and charged by law to finance the development of the Brazilian electricity sector and the Federal Government's energy policies.

The Sectoral Charges are:

- Quotas of the Global Reversion Reserve (RGR)
- Fuel Consumption Account (CCC) quotas
- Electricity Services Inspection Fee (TFSEE)
- Apportionment of Proinfa costs
- Energy Development Account (CDE)
- The Transmission Charges are:
- Use of Basic Electricity Transmission Network Facilities
- Use of Connection Facilities
- Use of Distribution Facilities
- Transport of Electricity from Itaipu
- National System Operator (ONS)

The Use of Basic Transmission Network Facilities is the revenue owed to all the electricity transmission companies that make up the Basic Network (the national interconnected system made up of transmission lines that carry electricity at a voltage equal to or greater than 230 kV) and which is paid by all generation and distribution companies, as well as large consumers (free consumers) who use the Basic Network directly, through a tariff for the use of the transmission systems - TUST.

The Use of Connection Facilities is the amount owed by electricity distribution companies that use transmission lines connected to the Basic Grid.

The Use of Distribution Facilities is the price paid by those who use the electricity networks owned by the distribution concession companies. This usually involves a generating company connected directly to the distribution company or a large energy consumer, such as steel companies (Arcelor), mining companies (Companhia Vale do Rio Doce) or petrochemical companies (Petrobrás). The price is determined by ANEEL through the Tarifa de Uso dos Sistemas Elétricos de Distribuição - TUSD.

Itaipu Electricity Transport is the cost paid by the electricity distribution companies that purchase quotas of electricity produced by the Itaipu Hydroelectric Power Plant, to reimburse the Operation and Maintenance costs of the direct current transmission networks used to take the power from the plant to the consumer market.

The transmission charge known as the National System Operator refers to the reimbursement of part of the administration and operating costs of the ONS (the entity responsible for operating and coordinating the Basic Grid) by all generation, transmission and distribution companies, as well as large consumers (free consumers) connected to the Basic Grid.

Parcel B are the amounts needed to cover the costs of personnel, material and other activities directly linked to the operation and maintenance of distribution services, as well as the costs of depreciation and remuneration of the investments made by the company to provide the service. These costs are identified as manageable costs because the concessionaire is fully capable of managing them directly and they have been agreed as components of "Parcel B" of the company's Annual Revenue Requirement. Parcel "B" can also be divided into three groups:

- Operating and Maintenance Expenses
- Capital Expenditure
- Other Expenses

Operation and Maintenance Expenses is the portion of revenue intended to cover costs directly linked to the provision of the electricity distribution service, such as personnel, materials, third-party services and other expenses. ANEEL does not recognise in the company's tariffs any costs that are not related to the provision of the service or that are not relevant to its geographical concession area. ANEEL determines that the price charged to the consumer should be sufficient to guarantee the entire operation of the concession company, i.e. it stipulates sufficient revenue to pay for everything from the company's president to the technicians in charge of maintaining the distribution system, as well as the administrative, accounting and judicial structure needed to keep the company providing an adequate service.

Capital Expenditure basically involves the Depreciation Quota and Capital Remuneration. The Depreciation Quota is the portion of revenue needed to build up the financial resources earmarked for restoring the investments made prudently to provide the electricity service at the end of its useful life. The purpose of this quota is to ensure, for example, that at the end of the useful life of a pole, electricity meter or transformer, the concessionaire has the revenue to replace the equipment with new equipment, maintaining the quality of the service provided.

The Remuneration of Capital is the portion of revenue necessary to promote an adequate return on the capital invested in the provision of the electricity service, i.e. the rate of return calculated as a rate on a Regulatory Remuneration Base.

In practice, ANEEL calculates a rate of return (profit) of approximately seven and a half per cent (7.5% = WACC (Weighted Average Capital Costs)) on the investments made by the distribution concessionaire and accepted by ANEEL as prudent and necessary for the provision of the service.

"Parcel B" also includes investments in Research and Development (R&D) and Energy Efficiency, and PIS/COFINS expenses.

Investments in Research and Development and Energy Efficiency are mandatory annual investments of at least 0.75% (seventy-five hundredths of a percent) of the company's net operating revenue in research and development in the electricity sector and at least 0.25% (twenty-five hundredths of a percent) in energy efficiency programmes, aimed at the end use of energy due to an obligation to
determined by Law No. 9,991 of July 2000.

The purpose of spending on research and development is to increase the quality of the service provided by electricity distribution companies, guaranteeing greater reliability in electricity systems, fewer interruptions in supply, less theft of electricity, etc.

The aim of spending on Energy Efficiency is to try to save electricity, for example by exchanging old appliances that consume a lot of energy for new, more economical equipment. According to the Tariff Regulation Procedures (Proret) [3] the regulatory costs allocated to the TUSD are defined in the tariff adjustment and revision processes. In these processes, cost functions are revised taking into account the distributor's costs related to energy transport, various charges and process losses. Figure 4 [3] shows the cost functions used to create the TUSD associated with one of the tariffs that are applied. Figure 4 also shows an important feature: the TUSD does not take into account the cost of energy; this value is only computed in the TE, which will be presented below.

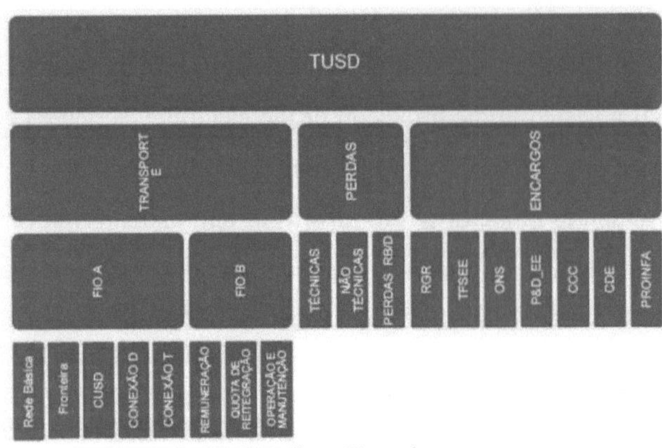

Figure 4- TUSD Cost Functions.

Source: Proret, submodule 7.1, 2011, p8

7.3. ENERGY TARIFF (TE)

As with TUSD, Proret establishes that the regulatory costs allocated to TE are defined in the tariff readjustment or revision process. In these processes, the cost functions are revised taking into account the distributor's costs related to charges, transport, losses and energy. Figure 5 [13] shows the TE cost functions.

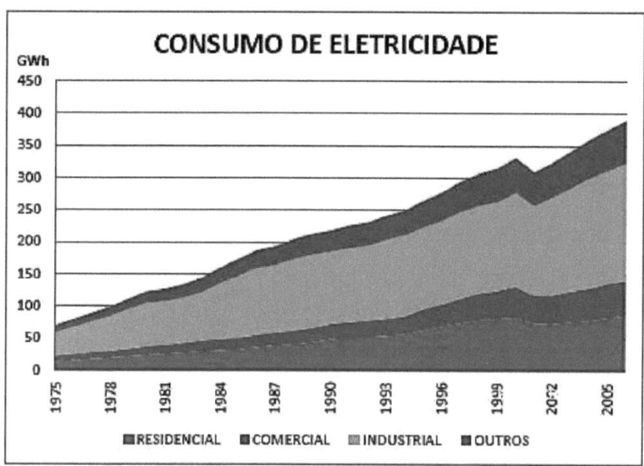

Figure 5 - Electricity consumption over the years

Source: Ministry of Mines and Energy, 2007

CHAPTER 8

NEW TARIFF STRUCTURE

Since 1982 and the publication of the first TUSD in 1999, few changes have been made to the tariff structure. On the other hand, the electricity sector has grown a lot, as can be seen in Figure 5 [12], which shows the growth in electricity consumption over the years in Brazil.

This growth ended up generating not only an imbalance in the tariff system, but also a lack of incentives for energy efficiency in installations.

8.1. INTERMEDIATE TARIFF STATION

According to the Technical Note published by ANEEL [13], studies suggest that the hourly rates for group A should be maintained, and that a third rate for group B should be created. This new post is intended to shift loads to potential peak hours, i.e. times when the electricity system is idle. The new post, applied only to the white tariff modality, will be called the intermediate tariff post or intermediate period. Following the same pattern as Figure 3, Figure 6 [15] shows that the intermediate period (IN) is defined as the period formed by the hour immediately before and the hour immediately after the peak period.

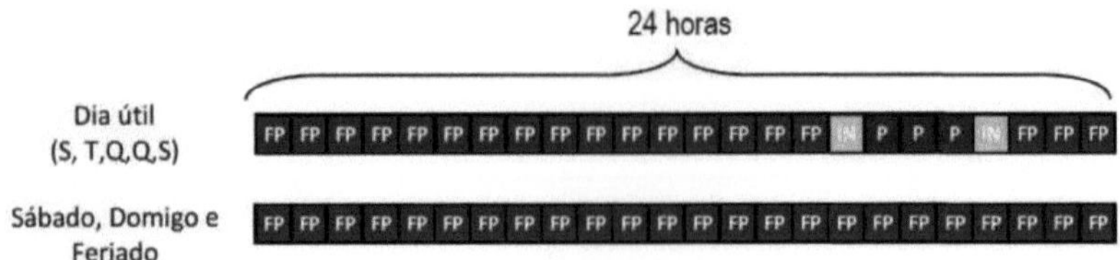

Figure 6- Tariff points for group B

Source: ANEEL, Technical note n°311/2011, 2011, p14

The off-peak period will now be defined as the period made up of the hours complementary to the peak and intermediate periods.

7.3. WHITE MODE

With the creation of the intermediate tariff station, the white modality was created, characterised by hourly signalling in the monomial energy tariff. It takes into account all three tariff periods and the total consumption made in each of these periods. This new modality, regulated in 2012 by ANEEL [15], is intended to be advantageous for consumers with the flexibility to change their consumption habits, and to help the electricity system to have a lower load during peak load times.

This new modality will be optional for all Group B consumers except for public lighting and low-income markets, which will have to remain in the conventional modality because they are not consumers with the flexibility to change their consumption pattern [3].

Since the white mode needs to record the value of consumption at each tariff station, it is necessary to replace the energy meters currently in use with more advanced meters. This change creates new measurement and monitoring possibilities that will be used in the future to update the tariff system.

7.4. SEASONAL SIGNALLING

Today's hourly system, published in 1982, is no longer sufficient to ensure that the energy market functions fairly. So the new model proposed for 2014 addresses this issue drastically. The proposal presented and accepted is for seasonal signalling to be replaced by the use of an economic signal marked by the so-called tariff flags. These will aim to indicate the cost of energy that will be applied in the next measurement cycle.

7.5. BANDS

The flag system is represented by three tariff flags [3]: green, yellow and red. Each flag represents the cost of energy to be applied in the next measurement cycle.

The triggering of each flag will be signalled monthly by ANEEL, according to the values of the Marginal Operating Cost (CMO) and the System Services Charge for Energy Security (ESS_SE).

For each month, the National Electricity System Operator (ONS) must estimate and publish the ESS_SE value using the Generation Cost for Energy Security (CGSE) and Projected Load (CP) values, according to Equation 14, and the estimated value for the CMO by submarket.

$$ESS_{SE} = \frac{CGSE}{CP} \qquad (14)$$

The three flags and their activation system are shown below.

7.5.1. Green

It corresponds to the financial economic equilibrium energy tariff and does not entail an additional cost to energy consumption. It will be used in the months when the sum of the CMO and ESS_SE values is less than R$100.00/MWh.

7.5.2. Yellow

This corresponds to an increase of R$15.00/MWh on the cost of energy during the financial economic balance. It will be used in the months when the sum of the CMO and ESS_SE values is between R$100.00 and R$200.00/MWh.

7.5.3. Red

This corresponds to an increase of R$30.00/MWh on the cost of energy during the financial economic balance. It will be used in the months when the sum of the CMO and ESS_SE values exceeds R$200.00/MWh.

Table 4 shows the approved energy tariffs for Elektro Eletricidade e Serviços S/A [17]. The new layout of the approved tariffs shows that the time-seasonality brought about by the flags is being attributed to all modalities.

Table 4 - Application tariffs for Elektro in the new model

ANNEX I - APPLICATION RATES

LOW VOLTAGE TARIFFS

TABLE A - CONVENTIONAL TARIFF MODE - ELEKTRO					
SUBGROUP/CLASS/SUBCLASS	TUSD	TE			
		TE	GREEN FLAG	YELLOW FLAG	RED FLAG
	RS/MWh	RS/MWh	RS/MWh	R$./MWh	RS/MWh
BI-RESIDENTIAL	213,03	139,14	139,14	154,14	169,14
B1 - LOW INCOME RESIDENTIAL	199.99	139,14	139,14	154,14	169,14
B2-RURAL	134.82	87,63	87.63	102,63	117,63
B2 - RURAL ELECTRIFICATION CO-OPERATIVE	108,58	70,57	70,57	85,57	100,57
B2 - PUBLIC IRRIGATION SERVICE	132.49	83.90	83.90	98,90	113,90
B3 - OTHER CLASSES	215,11	139,81	139,81	154.81	169,81
B4 - STREET LIGHTING					
B4a - Distribution Network	110,74	71.97	71,97	86,97	101,97
B4b - Lamp Bulb	121,60	79,03	79,03	94,03	109,03

TABLE B - WHITE TARIFF MODALITY																
SUBGROUP/CLASS/SUBCLASS	TUSD			TE												
	TIP	INTERMEDIATE	OUTSIDE	TIP				INTERMEDIATE				OUTSIDE				
				TE	GREEN FLAG	FLAG YELLOW	FLAG RED	TE	GREEN FLAG	YELLOW FLAG	BANDEIRA RED	TE	GREEN BANDEIRA	YELLOW FLAG	FLAG RED	
	RS/MWh	RS/MWh	RSMWh	RS/MWh	RSMWh	RS/MWh	RS/MWh	RS/MWh	R$^4Wh	R&MWh	RS/MWh	RS/MWh	RS/MWh	RS/MWh	RS/MWh	
BI - RESIDENTIAL	477,12	303,90	130,67	216,66	216.66	231,66	246.66	132,85	132,85	147,85	162,85	132,85	132,85	147,85	162,85	
B2- RURAL	340,13	215,02	89,91	134.52	134.52	149,52	164,52	82,48	82.48	97.48	112,48	82,48	82.48	97.48	112.48	
B2 - RURAL ELECTRIFICATION CO-OPERATIVE	273,91	173,16	72.40	108,33	108,33	123,33	138,33	66,42	66.42	81.42	96,42	66,42	66,42	81,42	96,42	
B2 - PUBLIC IRRIGATION SERVICE	328,70	207,79	86.89	130,00	130,00	145,00	160,00	79,71	79,71	94,71	109,71	79,71	79.71	94,71	109,71	
B3-OTHER CLASSES	612,69	385,07	157,45	214,61	214.61	229,61	244,61	131.59	131,59	146,59	161,59	131,59	131,59	146,59	161,59	

HIGH VOLTAGE TARIFFS

TABLE C - BLUE TARIFF MODE											
SUBGROUP	TUSD			TE							
	POMA	OUTSIDE	RS/MWh	POMA				OUTSIDE			
				TE	GREEN FLAG	FLAG YELLOW	FLAG RED	TE	GREEN FLAG	YELLOW FLAG	FLAG RED
	RS/kW	RS/kW		RS/MWh	RS/MWh	RS/MWh	RS/MWh	RS/MWh	RS/MWh	RS/MWh	RS/MWh
A2 (88 to 138kV)	10,27	3,86	25,01	215,60	215,60	230,60	245,60	132,20	132,20	147,20	162,20
A3(69kV)	17,94	4,59	32,02	219,46	219,46	234,46	249,46	134,56	134,56	149,56	164,56
A3a (30 to 44kV)	28,96	7,56	34,74	215,97	215,97	230,97	245,97	132,43	132,43	147,43	162,43
A4 (2.3 to 25kV)	36,21	9,45	34,74	215,97	215,97	230,97	245,97	132,43	132,43	147,43	162,43

TABLE F - GREEN TARIFF MODE											
SUBGROUP	TUSD			TE							
	RS/klV	TIP	OUTSIDE	TIP				OUTSIDE			
				TE	FLAG GREEN	YELLOW FLAG	FLAG RED	TE	FLAG GREEN	FLAG YELLOW	FLAG RED
		RS/MWh	RS/MWh	RS/MWh	RS/MWh	RS/MWh	RS/MWh	RS/MWh	RS/MWh	RS/MWh	RS/MWh
A3a (30 to 44kV)	7,52	727,49	34,56	215,97	215,97	230,97	245,97	132,43	132,43	147,43	162,43
A4 (2.3 to 25kV)	9,40	900,72	34,56	215,97	215,97	230,97	245,97	132,43	132,43	147,43	162,43

TABLE I - CONVENTIONAL TARIFF						
SUBGROUP	TUSD		TE			
			TE	GREEN FLAG	YELLOW FLAG	RED FLAG
	RS.kW	RS/MWh	RSMWh	RS/MWh	**RSMWh**	RS/MWh
A3a (30 to 44kV)	27,79	33,98	139,13	139,13	154,13	169,13
A4 (2.3 to 25kV)	34,74	33,98	139,13	139,13	154,13	169,13

Source: ANEEL, Homologatory Resolution No. 1,336, 2012, p.5

CHAPTER 9

UFAM UNITS ANALYSED

The Federal University of Amazonas (UFAM) has more than 20,000 students, 15 academic units and several administrative units, both on and off its campus in Manaus. There are also units located in the interior of the state, which we will analyse in the background.

UFAM is considered a major consumer of electricity, i.e. its energy demand and consumption is high compared to most consumers. In addition, the university is a major knowledge-producing centre, which requires almost continuous use of classrooms, laboratories - many of them with high-powered equipment - study rooms, lighting and so on.

It is a fact that this need for electricity puts the university at a different level in terms of electricity consumption. This generates the need for an in-depth energy efficiency study aimed at optimising electricity consumption.

Recently, UFAM has undergone constant changes in its physical structure. The new courses that were created did not have laboratories or classrooms to accommodate the new students who would be entering the university. It was therefore necessary to expand the institution's physical structure with the construction of new classroom buildings, new laboratories, rooms for new teachers, community centres, auditoriums, car parks and so on. All of these activities are closely linked to electricity consumption. A new laboratory will undoubtedly increase the final cost of electricity consumption. Imagine several of them.

What is astonishing, and the focus of this paper's analysis, is the lack of planning. There was no concern within UFAM about the increase in electricity consumption or the increase in electricity demand. There was no planning to calculate how much the increase in these variables would be and thus review the contract with the concessionaire. Now, all of this has caused a huge problem for UFAM - the absurd payment of fines for exceeding the contracted demand. Fines that practically equal the amount of the invoice. It's as if the university was paying the bill twice in one month.

The main aim of this work is to propose solutions to reverse this situation, using tools such as software to help choose the demand to be contracted and improvements to UFAM's electrical structure.

9.1. EXPERIMENTAL FARM

The UFAM Experimental Farm was created in 1974, in an area donated by SUFRAMA, as part of the agricultural production sector linked to the Faculty of Agricultural Sciences. Due to a number of factors, including a lack of financial and logistical resources, between 1979 and 1980 its occupation process was interrupted and its management was taken over by the Agricultural District Support Centre Foundation - FUCADA. This management lasted until 1994, when the farm was again administered by UFAM. In 2000, this experimental area was transformed into a supplementary body of UFAM, linked to the Rector's Office. As a supplementary body, it supports field activities in research and extension projects and teaching activities within its area.

The farm's facilities are:

- Fish farming/aquaculture laboratory (70 square metres)
- Alternative energy sources laboratory (120 square metres)
- Engine Testing and Carbonisation Laboratory (96 square metres)
- Beekeeping/meliponiculture laboratory (60 square metres)
- Classrooms and Administration Building (220 square metres - 3 classrooms; 1 Administration room; 1 Reception room; and 2 Special Projects rooms)
- Annex to the Classroom Building (110 square metres - 2 bathrooms; 1 bedroom/suite; 1 room for the Pilot Cracking Plant for Biodiesel Testing)
- Cafeteria for 70 people and kitchen (280 square metres)
- Warehouse and Reception Room (120 square metres)
- Accommodation 1 (220 square metres - 10 bedrooms and 02 bathrooms)
- Accommodation 2 (420 square metres - 12 bedrooms and 02 bathrooms)
- Accommodation 3 (180 square metres - 2 blocks with 04 bedrooms/suites each)

- Accommodation 4 (30 square metres - 02 bedrooms, living room, kitchen and bathroom)
- Accommodation 5 (30 square metres - 02 bedrooms, living room, kitchen and bathroom)
- Accommodation 6 (30 square metres - 02 bedrooms, living room, kitchen and bathroom)
- Laying aviaries 1 (70 square metres)
- Laying aviaries 2 (70 square metres)
- Intensive cutting aviaries (600 square metres)
- Semi-intensive broiler aviaries (600 square metres)
- Pigsty (1 with 100 square metres)
- Corralama and Brete (1000 square metres)
- Minhocariums (60 square metres)
- Feed mill (80 square metres)
- Sawmill and joinery (200 square metres)
- Garage/workshop building (135 square metres)
- Hydro-meteorological station (80 square metres - fenced space with an open-air screen)

Main activities carried out

- Teaching:

Around 30 subjects taught by professors from the University:

1. Faculty of Agricultural Sciences
2. Institute of Biological Sciences
3. Faculty of Technology

- Research:

1. Institutional projects: PIBIC, Master's and Doctorate.
2. Aquaculture and Fish Farming Laboratories and Alternative Energy Sources (Investments of around R$ 700,000.00)
3. Physical restructuring: refurbishment and construction of accommodation,

sheds, stables and others. Acquisition of equipment (investments of around R$200,000.00)

- Extension:

1. Courses and mini-courses given to farmers, the university community and the community in general.

9.2. ARTS CENTRE

For more than two decades, the Arts Centre of the Federal University of Amazonas has been conquering its place in the Amazonian community, contributing to the formation and training of individuals to develop their artistic thinking and doing.

This involvement began when the Joaquim Franco Conservatory was transferred from the state government to the University of Amazonas in 1968. A few years later, in 1987 to be precise, the conservatory became the Arts Sector and in the 1990s it was renamed the Arts Centre of the University of Amazonas.

During this historical process, the Arts Centre not only changed its name, but evolved and diversified its specific activities considerably, as a university supplementary body, provided for in the University Statute, focused on the field of the arts.

Its main objectives include acting as a university extension body to promote teaching, research and systematic artistic production within the university, as well as making an important contribution to the formation and development of citizens in society, by promoting activities at different levels: playful, expressive and pedagogical.

In this way, the Arts Centre is committed to disseminating artistic knowledge, providing the community with access to quality courses so that people can produce and express themselves through art, whether for leisure or cultural interest , or go on to higher education courses with a view to becoming professionals.

9.3. CENTRAL LIBRARY

The Central Library of the Federal University of Amazonas was set up on 12 September 1974. It is a Supplementary Body with several offices and, since January

1992, it has occupied the ground floor of the Sectorial Library building of the Faculty of Health Sciences, where it has partially remained since April 2002. The other sectors are temporarily operating on the University Campus, on the 2nd floor of the Campus Sector Library, in a building belonging to the Faculty of Social Studies.

Considered a provider of knowledge and fundamental support for the pillars of teaching, research and extension, it has a centralised organisation that promotes the loan of all the bibliographic and audiovisual material needed and demanded by the university community, as well as the transfer to the sector libraries according to the requests made, through the University Library System.

9.4. FACULTY OF DENTISTRY

The dentistry course at the Federal University of Amazonas was created in 1966 by Resolution No. 4069-A of June 1962 and recognised in 1973 by Decree No. 71768 of 26 January 1973. It first operated in the Nilo Peçanha school building and was later transferred to the Aparecida neighbourhood when the Pharmacy and Dentistry course at the University of Amazonas began, where the theoretical classes were taught, with the Araújo Lima Outpatient Clinic remaining as the environment where a consulting room was initially set up and practical classes were taught there.

The year 1980 brought changes for the Dentistry course, which was split off from the Pharmacy course and moved to the current building, located in the Praça XIV de Janeiro neighbourhood.

9.5. FACULTY OF NURSING

The Manaus School of Nursing is an academic unit that offers undergraduate and postgraduate courses, specialising in Nursing/Health and a Master's degree in Nursing. The undergraduate programme is open to high school graduates and the postgraduate programme is open to graduates with a degree in Nursing or another higher education course. The library is open to the university community and society in general.

9.6. MANAUS UNIVERSITY CAMPUS

The area of the university campus - 6.7 million square metres - makes it the third largest green fragment in an urban area in the world and the first in the country. It is home to various species of fauna - such as sloths, pacas and collared sauins - and flora, amidst a large area of virgin forest. The built area corresponds to around 35% of the original architectural project by Severiano Mário Porto, which earned him an honourable mention in 1987 from the Institute of Architects of Brazil (IAB/RJ).

The University's administrative structure is made up of the highest administrative body, the Rectorate, followed by the Pro-Rectories and Supplementary Bodies, each of which has various sectors as described below:

Rectory: made up of the Cabinet, the Executive Board, the General Secretariat of the Higher Councils, the Representation in Brasilia, the Communications Office, the Legal Prosecutor's Office, the Internal Audit and the Management Committee;

Dean's Offices: divided into Undergraduate Education, Research and Postgraduate Education, Extension and Internalisation, Planning, Administration and Community Affairs;

It currently offers 102 undergraduate programmes and 40 stricto sensu postgraduate programmes accredited by Capes. There are a total of 32 Masters programmes and 8 Doctorates. At the Lato sensu postgraduate level, there are more than 30 courses on offer each year. As far as Extension is concerned, there are over 600 projects that directly benefit the population and 17 major extension programmes.

9.7. THE SOFTWARE FOR CALCULATING OPTIMUM DEMAND AND ITS APPLICABILITY: FINES FOR EXCEEDING DEMAND

This software consists of a spreadsheet developed in Microsoft Office Excel, which analysed a year's worth of electricity bills from various sectors at UFAM. The software was developed and adapted with the help of UFAM professors Lucas Carvalho Cordeiro and Rubem Cesar Rodrigues Souza.

The analysis to be carried out by the spreadsheet will use Excel's Solver tool. Solver is part of a package of programmes sometimes called hypothesis testing tools.

With Solver, you can find an ideal value (maximum or minimum) for a formula in a cell - called the target cell - according to restrictions, or limits, on the values of other formula cells in a spreadsheet. Solver works with a group of cells, called decision variables or simply variable cells, which take part in the calculation of the formulae in the goal and constraint cells. Solver adjusts the values in the decision variable cells to satisfy the limits on the constraint cells and produce the result you want for the objective cell.

Thus, by evaluating the billing history, the conditions for the cells were imposed, considering that there would only be a fine when the value exceeded the contracted value by 5 per cent (this is because high voltage is being considered).

It's worth noting that the amount of demand currently contracted by UFAM is 2000 kW and that the normal demand tariff is worth R$36.96/kW. The overrun tariff is worth twice as much, i.e. R$73.92/kW. In the end, the programme also calculated the annual cost of demand if the demand value were adopted. We will therefore make comparisons to show how much UFAM would save if it adopted the measures proposed in this work.

Table 5 below shows the historical energy demand consumption of the UFAM University Campus between June 2012 and May 2013.

Table 5 - UFAM Manaus Campus demand history

Month/Year	Demand Consumption (kW)
JUNE/2012	2380
JULY/2012	1968
AUGUST/2012	2472
SEPTEMBER/2012	3182
OCTOBER/2012	3417
NOVEMBER/2012	2472
DECEMBER/2012	2952
JANUARY/2013	2956
FEBRUARY/2013	2846
MARCH/2013	3168
APRIL/2013	3129
MAY/2013	2505

It is immediately apparent from looking at this table that the contracted demand figure of 2000 kW is totally inadequate and out of date, the result of a total lack of planning on the part of UFAM's electricity sector. Refurbishments, the construction of new buildings and laboratories that have increased energy demand have been ignored in recent years. Below is Table 6, which shows the electricity demand consumption of UFAM's Nursing building, which is outside the University Campus, whose contracted demand value, until November 2012, was zero, i.e. all the demand was paid for as an overrun. Currently, the contracted value shown on the invoice is 138kW.

Table 6 shows the history of UFAM's experimental farm, located at BR-174, km 38. This unit had no contracted demand until December 2012, and even today it does not have an adequate power factor, which is between 0.82 and 0.88. Today the contracted demand is 76kW

Table 6 - Historical demand for UFAM's experimental farm

Month/Year	Consumed Demand (kW)
JUNE/2012	47
JULY/2012	56
AUGUST/2012	54
SEPTEMBER/2012	56
OCTOBER/2012	67
NOVEMBER/2012	60
DECEMBER/2012	55
JANUARY/2013	76
FEBRUARY/2013	50
MARCH/2013	57
APRIL/2013	49
MAY/2013	69

Table 7 shows the case of the UFAM Arts Centre, located at Rua Monsenhor Coutinho, 724 - Centro. This case is the opposite of those already shown, i.e. initially there was no contracted demand, but the one that was subsequently contracted, of 52kW, is considered excessive, making it an unnecessary expense for UFAM.

From now on, we'll resume the application of the software. The spreadsheet that will be used has the demand history of various sectors of the Federal University of

Amazonas, and with the help of the solver we will analyse the history of these demands in order to choose the optimum demand to be contracted.

Table 7 - History of demand at the UFAM Arts Centre

Month/Year	Consumed Demand (kW)
JUNE/2012	31
JULY/2012	30
AUGUST/2012	25
SEPTEMBER/2012	32
OCTOBER/2012	30
NOVEMBER/2012	32
DECEMBER/2012	28
JANUARY/2013	18
FEBRUARY/2013	21
MARCH/2013	22
APRIL/2013	24
MAY/2013	18

CHAPTER 10

APPLICATION OF OPTIMAL DEMAND CALCULATION SOFTWARE

With UFAM's electricity bill history from June 2012 to May 2013, the consumption behaviour of the units was surveyed. The units analysed were:

Figura 1: Experimental farm

Figura 2: Arts Centre

Figura 3: Central Library

Figura 4: Faculty of Dentistry

Figura 5: Faculty of Nursing

Figura 6: University Campus

The software was responsible for calculating the demand to be contracted and also the annual cost of demand, if the demand proposed by the software was contracted.

Therefore, we will show each unit analysed and make the appropriate comparisons.

This unit has a complex system of activities closely linked to electricity consumption. The survey was therefore carried out within the software, with the data already shown in table 6. The result is shown in figure 7 below:

Standard tariff (R$/kW)	36,96	NOTE: This amount is shown on the electricity bill.
Overrun tariff (R$/kW)	73,92	NOTE: This amount is shown on the electricity bill.
Demand to be contracted (kW)	**72**	NOTE: This value is generated by the programme.

Figura 7: Determining the demand to be contracted for the UFAM experimental farm

The farm's current contracted demand is 76kW, which generates an annual demand cost of R$33,707.52. If the value calculated by the programme were used, the cost would be R$32,102.40, which would generate savings of R$1,605.12 per year, considering only the farm.

This is a relatively small amount, but any savings are welcome. What's more, this money could be invested in improving the unit's power factor, or in improving the farm's basic services.

With the history of the Arts Centre already shown in table 7, we have the result of the survey carried out by the software shown in figure 8 below:

Standard tariff (R$/kW)	36,96	NOTE: This amount is shown on the electricity bill.
Overrun tariff (R$/kW)	73,92	NOTE: This amount is shown on the electricity bill.
Demand to be contracted (kW)	30	NOTE: This value is generated by the programme.

Figura 8: Determining the demand to be contracted for the Arts Centre

At the moment, the demand contracted by the Arts Centre is 52kW, so the annual cost of demand is R$23,063.04. If the value calculated by the programme were used, the annual cost would be R$13,516.80, which would generate savings of R$9,547.24 in one year.

This is now a relatively significant amount in terms of savings, which could be applied, among other things, to repairs and maintenance at the Arts Centre.Table 8 below shows the history of demand measured by the Central Library.

Table 8 - Historical demand for the UFAM Central Library

MONTH/YEAR	MEASURED DEMAND (kW)
JUNE/2012	52
JULY/2012	50
AUGUST/2012	63
SEPTEMBER/2012	67
OCTOBER/2012	72
NOVEMBER/2012	75
DECEMBER/2012	62
JANUARY/2013	58
FEBRUARY/2013	65
MARCH/2013	65
APRIL/2013	63
MAY/2013	50

Today, the demand contracted by the Central Library is 190kW, generating an annual demand cost of R$84,268.80. The value calculated by the programme, 71kW, shown in figure 9 below, would generate an annual cost of R$31,680.00, saving, therefore, R$52,588.80. A high figure compared to those already presented here, which could be used to renew the library's collection, for example.

Standard tariff (R$/kW)	36,96	**NOTE: This amount is shown on the electricity bill.**
Overrun tariff (R$/kW)	73,92	**NOTE: This amount is shown on the electricity bill.**
Demand to be contracted (kW)	**71**	NOTE: This value is generated by the programme.

Figura 9: Determining the demand to be contracted by the Central Library

Table 9 shows the demand history of the UFAM Faculty of Dentistry, over the same period analysed for the other units.

The demand contracted by the School of Dentistry is 350kW.

Therefore, the annual cost of demand is R$ 155,232.00. Now, if the value calculated by the programme were used, which is shown in figure 10 and is 298kW, the annual cost of demand would be R$132,211.20. This would generate savings of R$23,020.80, which is a reasonable amount of savings, enough for internal improvements and repairs at the Faculty of Dentistry.

Table 9 - History of demand from the Faculty of Dentistry

MONTH/YEAR	MEASURED DEMAND (kW)
JUNE/2012	254
JULY/2012	213
AUGUST/2012	220
SEPTEMBER/2012	305
OCTOBER/2012	313
NOVEMBER/2012	306
DECEMBER/2012	292
JANUARY/2013	275
FEBRUARY/2013	262
MARCH/2013	291
APRIL/2013	259
MAY/2013	219

Standard tariff (R$/kW)	36,96	**NOTE: This amount is shown on the electricity bill.**
Overrun tariff (R$/kW)	73,92	**NOTE: This amount is shown on the electricity bill.**
Demand to be contracted (kW)	**298**	NOTE: This value is generated by the programme.

Figura 10: Determining the demand to be contracted by the Faculty of Dentistry

The Faculty of Nursing's demand history is shown in Table 10 below.

Table 10 - Historical demand from the Faculty of Nursing

MONTH/YEAR	MEASURED DEMAND (kW)
JUNE/2012	59
JULY/2012	57
AUGUST/2012	71
SEPTEMBER/2012	73
OCTOBER/2012	74
NOVEMBER/2012	78
DECEMBER/2012	55
JANUARY/2013	56
FEBRUARY/2013	102
MARCH/2013	66
APRIL/2013	53
MAY/2013	122

With this history, the software calculated an optimum demand of 116 kW, as shown in figure 11.

Standard tariff (R$/kW)	36,96	**NOTE: This amount is shown on the electricity bill.**
Overrun tariff (R$/kW)	73,92	**NOTE: This amount is shown on the electricity bill.**
Demand to be contracted (kW)	**116**	NOTE: This value is generated by the programme.

Figure 11: Determining the demand to be contracted by the Faculty of Nursing

Now, since the amount of demand contracted by the Faculty of

This gives us an annual cost of R$61,205.76. By adopting 116kW, the cost would fall to R$ 51,532.80, generating annual savings of R$ 9,672.96.

In the next section, the last case, which is also the most damning, will be assessed: the Manaus University Campus.

The case of the University Campus is different from the previous ones, because in the other units, the contracted demand was higher than the amount needed. In this case, the Campus' contracted demand, which is 2000kW, needs to be increased. As shown in Table 5, only in July 2012 was this figure not exceeded. The fines that UFAM pays for this are heavy, and can exceed 100,000 reais in one month. The amounts are shown below in table 11. It's worth remembering that any excess demand is paid at double the normal rate.

Table 11 - History of UFAM Campus Fines

MONTH/YEAR	DEMAND (KW)	EXCESS (KW)	COST (R$)	FINE (R$)
JUN/12	2380	380	87.964,80	28.089,60
JUL/12	1968	Zero	73.920,00	-
AUG/12	2472	472	91.365,12	34.890,24
SEPT/12	3182	1182	117.606,72	87.373,44
OCT/12	3417	1417	126.292,32	104.744,64
NOV/12	2472	472	91.365,12	34.890,24
DEC/12	2952	952	109.105,92	70.371,84
JAN/13	2956	956	109.253,76	70.667,52
FEB/13	2846	846	105.188,16	62.536,32
MAR/13	3168	1168	117.089,28	86.338,56
APR/13	3129	1129	115.647,84	83.455,68
MAY/13	2505	505	92.584,80	37.329,60
TOTAL		**9479**	1.237.383,84	700.687,68

The normal annual cost, i.e. without demand overruns during the entire period analysed, would be R$887,040.00. However, due to the overruns that occurred and adding the fines, we have a total amount paid by UFAM of R$1,938,071.52, which is 218.5% more. There is certainly a serious flaw here on the part of the institution, as the difference between the amounts is glaring.

By analysing the history of the campus, the software determined an optimum demand for the unit, as shown in figure 12 below.

Standard tariff (R$/kW)	36,96	**NOTE: This amount is shown on the electricity bill.**
Overrun tariff (R$/kW)	73,92	**NOTE: This amount is shown on the electricity bill.**
Demand to be contracted (kW)	**3.310**	NOTE: This value is generated by the programme.

Figura 12: Determining the demand to be contracted by the Manaus University Campus

With a demand value of 3,310 kW, the annual cost would be R$1,389,991.68 , which would generate savings of R$548,079.84 over the period analysed. This figure is highly significant, as the Manaus Campus needs a lot of investment and these savings would be fundamental for UFAM.

CHAPTER 11

FINAL CONSIDERATIONS AND SUGGESTIONS FOR IMPROVEMENT.

The Federal University of Amazonas is undoubtedly a major centre for the development of knowledge in the state of Amazonas. Its teaching, research and extension activities are extremely important for the development of local and national society. The university's capital is the brain. The university's role is to develop it. UFAM must develop creative minds to solve the future problems of society and humanity.

It is for this purpose that this final section suggests improvements. Although the case study already points to a solution that generates significant savings for UFAM, there are still other solutions within Electrical Engineering that need to be made.

To this end, here are some suggestions for reducing electricity consumption at the University.

- **AIR CONDITIONING**

Keep windows and doors closed to prevent air from entering.

Limit use of the device to occupied premises only.

Avoid sunlight in the air-conditioned room, as this will increase the heat load on the air conditioner.

Clean the appliance's filter as often as recommended by the manufacturer to prevent dirt from impairing its performance.

Keep the condenser air inlet free.

In winter or on cold days switch off the central or individual air-conditioning units and only use the ventilation system.

Whenever possible, use temperature control (thermostat) sectorised by room.

Thermal comfort is a combination of temperature and humidity, with the recommended range being between 20 and 22°C in winter and 23 to 25°C in summer, with 50 to 60 per cent relative humidity. Maximum cold is not always the best comfort solution. Air conditioning is responsible for the largest share of energy consumed by the university. Increasing the temperature by one degree will save the system around 7 per cent in electricity.

- **ELEVATORS**

Carry out awareness campaigns so that users prefer to use stairs for the first floors.

Locate the services with the highest footfall on the lower floors.

- **INFORMATION EQUIPMENT**

Switch off the computer at lunchtime

Programme switch-off for 11pm

- **LIGHTING**

Switch off lights in premises when they are not in use, such as meeting rooms, toilets, internal and external ornamental lighting.

Avoid switching on lamps during the day (use natural light whenever possible).

Group together sectors that need similar light intensities.

When cleaning large areas, switch on only the necessary lights.

Keep lamps and luminaires clean to allow maximum light reflection.

In outdoor spaces, reduce lighting in circulation areas, car parks and garages, where possible and without prejudice to safety.

Preferably use open luminaires, removing acrylic where possible, which makes it possible to reduce the number of lamps by up to 50 per cent without losing the quality of lighting.

Replace incandescent light bulbs with compact fluorescents.

In gardens, outdoor car parks and leisure areas, give preference to sodium vapour lamps.

Use electronic reactors with a high power factor.

Use presence sensors in little-used areas.

Lower the luminaires when the ceiling height is high, thereby reducing the total power required.

Design localised lighting when the activity requires it, proportionally reducing the general lighting in the room.

Install independent circuits and sensors with photocells in the areas near the windows, which automatically adjust the lighting levels needed to complement the natural light.

Walls, floors and ceilings should be painted in light colours that require less artificial lighting. Reducing the lighting load consequently reduces the thermal load for the air conditioning system.

In addition, it is worth remembering that the substation at the UFAM University Campus is still at the 13.8kV voltage level. Now, if UFAM migrated to the 69kV voltage level, whose tariff is cheaper, there would be even greater savings on electricity bills. Therefore, according to table 1, the University would fall into sub-group A3, where its tariff would be Blue.

For this tariff modality, the demand prices are R$263.42 per MWh (energy) and R$20.66 per kW (demand) during peak hours and R$160.19 per MWh (energy) and R$5.66 per kW (demand) during off-peak hours. Currently, UFAM pays R$ 38.54 per kW of

demand, i.e. the value of 69kV is almost 7 times lower off-peak, and even on-peak it is 46 per cent cheaper.

Still on the subject of reducing the final value of electricity bills, we have the case of the low power factor of some units such as: the Experimental Farm, the Faculty of Dentistry, the Faculty of Nursing and the Central Library. These units need to stop paying for excess reactive power.

A low power factor indicates that your UFAM is not making good use of its energy. In this case, the following situations can occur:

1. Increase in the installation's internal electrical losses
2. Voltage drop in the installation
3. Reduced utilisation of transformer capacity
4. Heated conductors

Low Power Factor can be corrected by: The correct sizing of motors and equipment, the permanent use of high Power Factor reactors, the installation of capacitors or capacitor banks where necessary (preferably close to the load), and the installation of synchronous motors in parallel with the load.

When the Power Factor is corrected and raised to 0.92 or more, energy is used more correctly and economically. That's because

1. The surcharge on electricity bills disappears
2. Improves the utilisation of electrical energy to generate useful work
3. Reduce voltage variations (oscillations)
4. Improves equipment utilisation with less consumption
5. Increases equipment lifespan
6. The conductors become less heated, reducing electricity losses in the installation.
7. Due to the release of load, the capacity of the transformers is better utilised.

Finally, the university should rethink the issue of energy planning . UFAM continues to expand, opening new courses, hiring new professors and demanding more and more electricity. This should always be under constant scrutiny. One suggestion would be for the Electrical Engineering faculty and the students to be jointly responsible for this monitoring. A university that has a Faculty of Technology should be the first to come up with solutions that use its teaching staff and students to solve problems related to saving electricity.

It is also suggested that there should be greater involvement and interaction between the university's various courses, such as architecture and engineering, in order to raise everyone's awareness of the increasing use of more energy-efficient electrical equipment. Next, the university should move towards the construction of green buildings. Green buildings are buildings that follow certain parameters and have a special concern for the environment in which they are located and for the correct use of the natural resources necessary for their operation and the correct disposal of the waste generated by this use. Thus, the concern for efficiency and quality is always aimed at minimising environmental impact. The benefits of such buildings are:

1. Energy efficiency;
2. Rational use of water;
3. Improved use of materials and resources;
4. Internal environmental quality;
5. Sustainable space;
6. Innovations and technologies;

I think this is important because of the privileged location of the Federal University of Amazonas, in the heart of the Amazon, where forest conservation is defended the world over.

One way of making this happen would be to hold an Engineering and Architecture week at UFAM's Faculty of Technology, in order to showcase the various solutions that the courses propose for improving the university's energy efficiency.

CONCLUSION

It was possible to conclude from this work that the structure of the electricity tariff is still evolving. Based on very basic concepts, it governs consumer behaviour and ensures a satisfactory balance between consumers and distributors. It can be seen that even with the addition of a new model with a slightly different philosophy, the basis of the calculations and the rules are the same.

Despite the improvements introduced by the new system that distributes energy costs quickly, the study of energy pricing needs to evolve in order to be able to make reliable predictions about future costs. This is because tariffs are dependent on water availability and reservoir management, which creates a probability problem that requires new tools.

With regard to the type of programme created, it can be concluded that its simplicity, both in terms of programming and operation, make it very attractive and adaptable. Its dynamism could favour not only the study of electricity demand contracting, but also the dissemination of knowledge held by a few and ignored by many.

Finally, this work presents in a very simple way some of the characteristics, needs and problems related to the electricity pricing model applied in Brazil and proposes solutions for improvements to reduce the enormous costs of UFAM's electricity bills.

REFERENCES

[1] SASSI. P. M, LEITE. A. F, CARMEIS. D. W, CARVALHO. C. B, REZENDE. M. R, RAMOS. D. S, JANNUZZI. G. M, **"Development of New Tariffs Horosazonal and Special Tariffs for Interruptible Electricity Supplies";** Final Report. 2002.

[2] MARQUES.M.C.S, HADDAD.J, MARTINS.A.R.S, **"Conservação de Energia. Energy Efficiency of Equipment and Installations",** 2006.

[3] **Tariff Structure for Distribution Concessionaires**; Submodule 7.1; Tariff Regulation Procedures: 2011, Brazil, 2011.

[4] **Law No. 9.428**, Brazil, 1996.

[5] **Normative Resolution No. 414**, ANEEL, 2010,

[6] **Elektro Energy Efficiency Manuals. Industrial Segment**, Elektro - Eletricidade e Serviços S.A ,

[7] **Decree No. 62.724**, Brazil, 1968.

[8] VIEIRA Junior.J.C, SEL0437 - **Energy Efficiency**, 2012,

[9] **National Energy Balance**, Ministry of Mines and Energy, 2012.

[10] **Load Consolidation for 2008-2010**, ONS, 2007

[11] **Manual do Cliente Horossazonal**, Escelsa, 2004.

[12] **National Energy Balance**, Ministry of Mines and Energy, 2007.

[13] **Technical Note No. 361/2010-SRE/SRD/ANEEL**, 2010.

[14] **Technical Note No. 311/2011-SRE/SRD/ANEEL**, 2011.

[15] **Normative Resolution No. 502, ANEEL**, 2012.

Printed by Books on Demand GmbH, Norderstedt / Germany